Artificial Intelligentsia

An interdisciplinarian perspective

Bob Seeman

CyberCurb

Copyright © 2023 by Bob Seeman

All rights reserved.

No part of this book may be reproduced, or stored in a retrieval system, or transmitted in any form or by any means, electronic, mechanical, photocopying, recording, or otherwise, without express written permission of the publisher.

ISBN: 9798386740276 (Hardcover)
ISBN: 9798386740153 (Paperback)

Cover design by: CyberCurb

Publisher: CyberCurb, Vancouver

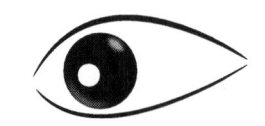

"The good news is that the computer has passed the Turing test. The bad news is that you have not."

About the Author

Bob Seeman is the Managing Partner of CyberCurb and a Director of the Cyber Future Foundation Canada, an international collaboration of industry, public agencies and academia to build a more trusted and secure internet. Bob is also a Mentor at the Rogers Cybersecure Catalyst.

Bob also has published *Who am I Not?*, *Power in Mistakes*, *On Trust*, and the foremost bitcoin-skeptic book, *The Coinmen*.

He is a California attorney, electrical engineer, and board director. Bob is a co-founder and former director of RIWI Corp., a public company that conducts data analytics, and has advised governments internationally on technology and business issues. Previously, he was Head of Strategy for Microsoft Network in London, and a technical consultant to the European Commission.

Bob previously practiced administrative law with an international law firm. He holds a Bachelor of Applied Science (Elec. Eng.) with Honours from the University of Toronto, a Master of Business Administration from EDHEC, and a Juris Doctor (J.D.) from the University of British Columbia.

For HAL

Table of Contents

The Interdisciplinarian Perspective	9
Consciousness	13
Alan Turing, mathematician (1912 – 1954)	14
Lawrence Krauss, theoretical physicist (1954 –)	16
Edsger Dijkstra, computer scientist (1930 – 2002)	18
Sebastian Thrun, entrepreneur (1967 –)	20
Alan Perlis, computer scientist (1922 – 1990)	22
Steve Polyak, computer scientist (1968 –)	24
Richard Dawkins, evolutionary biologist (1941 –)	26
Existential threat	29
Stephen Hawking, theoretical physicist (1942 – 2018)	30
Ray Kurzweil, computer scientist (1948 –)	32
Peter Thiel, entrepreneur (1967 –)	34
Eliezer Yudkowsky, writer (1979 –)	36
Nick Bostrom, philosopher (1973 –)	38
Louis B. Rosenberg, engineer (1969 –)	40
James Barrat, filmmaker (1960 –)	42
George Dyson, author (1953 –)	44
Benefits	47
Oren Etzioni, entrepreneur (1964 –)	48
Fei-Fei Li, computer scientist (1976 –)	50
David Gelernter, computer scientist (1955 –)	52

How it works   **55**

   Marvin Minsky, cognitive scientist, (1927 – 2016)   56

   Jaan Tallinn, investor (1972 –)   58

   Jensen Huang, business magnate (1963 –)   60

   Daniel Kahneman, psychologist and economist (1934 –)   62

   Daniel Crevier, entrepreneur (1947 –)   64

   John McCarthy, computer scientist (1927 2011)   66

Adapting   **69**

   Jaron Lanier, computer philosophy writer (1960 –)   70

   Peter Diamandis, entrepreneur (1961 –)   72

   Alan Kay, computer scientist (1940 –)   74

   Peter Watts, science fiction author (1958 –)   76

Acknowledgments   **83**

"Let's hold off on artificial intelligence until we figure out actual intelligence."

The Interdisciplinarian Perspective

With the release of ChatGPT at the end of 2022, artificial intelligence – or AI – became all the rage.

This book includes important quotations about AI by the Artificial Intelligentsia, people from many different academic disciplines who *also* understand *other* disciplines. The Artificial Intelligentsia have the Interdisciplinarian Perspective: they come, not just from technology backgrounds, but also from philosophy, art, language, writing, entertainment, astronomy, science, and business. The quotations illustrate how advances with AI will continue to result from the Artificial Intelligentsia who possess the Interdisciplinarian Perspective.

Many of the Interdisciplinarians quoted in this book disagree on many fundamental issues surrounding AI, including on the whether AI will improve the lives of humans – or destroy humans forever.

Each quotation is followed by a funny limerick – written by the human author of this book. AI does not yet have a funny bone. Rest assured that humans will be making funnier limericks than AI for a very long time.

"Artificial intelligence is when you receive your university degree but are still stupid."

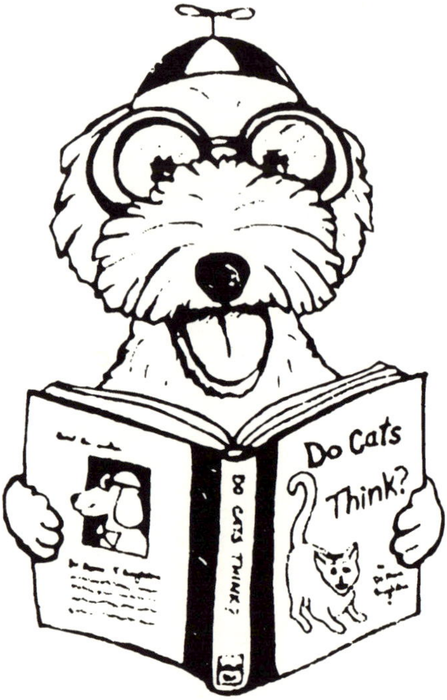
Do Cats Think?
DO CATS THINK?

Consciousness

> "A computer would deserve to be called intelligent if it could deceive a human into believing that it was human."

> "Instead of trying to produce a programme to simulate the adult mind, why not rather try to produce one which simulates the child's? If this were then subjected to an appropriate course of education one would obtain the adult brain."

Alan Turing, mathematician (1912 – 1954)

There once was a man quite alluring
Whose fame is forever enduring
He could break any code
That was then à la mode
And his name, yes you guessed it,
was Turing.

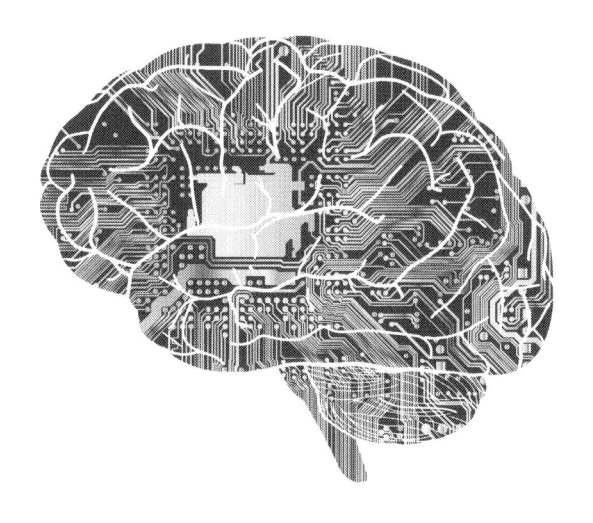

"There's an inverse relationship between what's known in a field and the number of books that are written about it."

We know very little about consciousness. It's generally accepted that activity in the brain is what generates the mental processes and our subjective experience of reality. But we know nothing about how it emerges. How bioelectrical signals give rise to the rich, boundless realm of thoughts, feelings, and sensations is completely unknown."

Lawrence Krauss, theoretical physicist
(1954 –)

The bell curve is called after Gauss
Just like waltzes are named after Strauss
He's incurred lots of wrath
But in physics and math
The name to remember is Krauss.

"The question of whether a computer can think is no more interesting than the question of whether a submarine can swim."

Edsger Dijkstra, computer scientist (1930 – 2002)

There once was a guy who was Dutch
Whom techies admire very much
Not self-promoting
A master at coding
Dijkstra, no one could touch.

"Nobody phrases it this way, but I think that artificial intelligence is almost a humanities discipline. It's really an attempt to understand human intelligence and human cognition."

Sebastian Thrun, entrepreneur (1967 –)

A giant computer nerd, Thrun
Considers AI to be fun
Like Tolkien and hobbits
He plays with his byte/bits
His input has hardly begun.

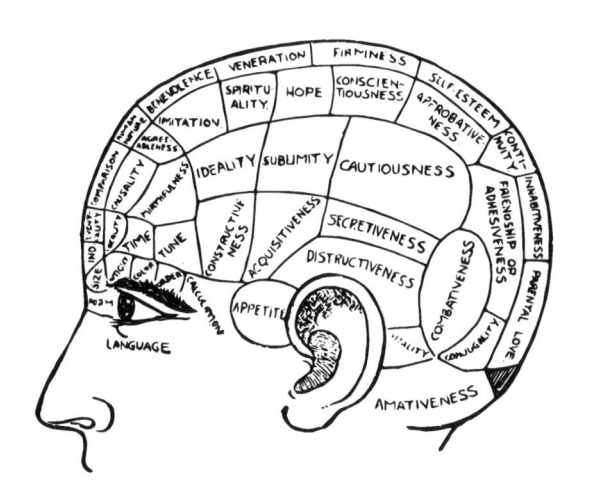

> *"A year spent in artificial intelligence is enough to make one believe in God."*
>
> Alan Perlis, computer scientist (1922 – 1990)

Alan Jay Perlis
Never is careless
He can program a language
While eating a sandwich
But Alan Jay Perlis is hairless.

"Before we work on artificial intelligence why don't we do something about natural stupidity?"

Steve Polyak, computer scientist (1968 –)

Was there ever a fellow like Steve
Whose deeds are quite hard to retrieve
'cause there's Stephens and Stevens
At odds and at evens
You don't know which name to believe.

"There is a popular cliché... which says that you cannot get out of computers any more than you put in. Other versions are that computers only do exactly what you tell them to, and that therefore computers are never creative. The cliché is true only in the crashingly trivial sense, the same sense in which Shakespeare never wrote anything except what his first schoolteacher taught him to write – words."

Richard Dawkins, evolutionary biologist
(1941 –)

There's a guy, Richard Dawkins by name
Who's acquired considerable fame
A huge nonbeliever
Works like a beaver
He's putting all others to shame.

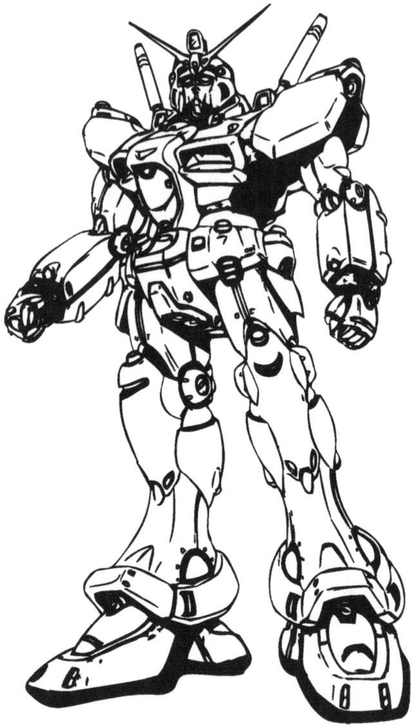

Existential threat

"The development of full artificial intelligence could spell the end of the human race. ... It would take off on its own, and re-design itself at an ever-increasing rate. Humans, who are limited by slow biological evolution, couldn't compete, and would be superseded."

"Everything that civilisation has to offer is a product of human intelligence; we cannot predict what we might achieve when this intelligence is magnified by the tools that AI may provide, but the eradication of war, disease, and poverty would be high on anyone's list. Success in creating AI would be the biggest event in human history. Unfortunately, it might also be the last."

Stephen Hawking, theoretical physicist
(1942 – 2018)

There once was a genius called Hawking
Who couldn't do walking or talking
His body disabled
His mind multi-fabled
His brain power awesome and shocking.

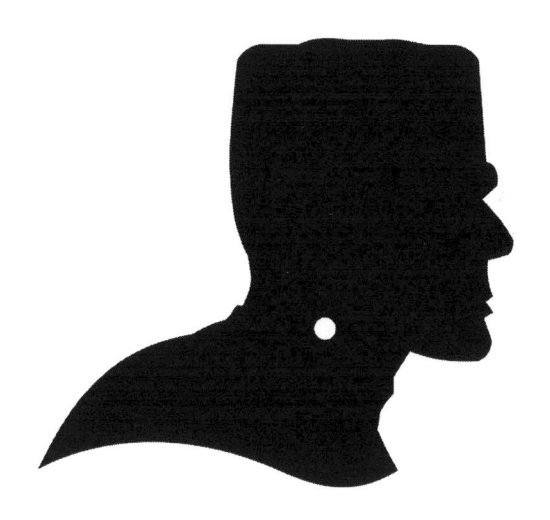

"Within a few decades, machine intelligence will surpass human intelligence, leading to The Singularity – technological change so rapid and profound it represents a rupture in the fabric of human history."

"With the increasingly important role of intelligent machines in all phases of our lives – military, medical, economic and financial, political – it is odd to keep reading articles with titles such as Whatever Happened to Artificial Intelligence? This is a phenomenon that Turing had predicted: that machine intelligence would become so pervasive, so comfortable, and so well integrated into our information-based economy that people would fail even to notice it."

Ray Kurzweil, computer scientist (1948 –)

Ray Kurzweil, a man with a vision
Believes in the need for precision
He has faith that machines
Will get rid of routines
His ideas invite some derision.

"People are spending way too much time thinking about climate change, way too little thinking about AI."

Peter Thiel, entrepreneur (1967 –)

There once was a smartie called Peter
Who some maybe thought was a cheater
He used Chat GPT
No matter the fee
And thought there was naught that was neater.

"The AI does not hate you, nor does it love you, but you are made out of atoms which it can use for something else."

Eliezer Yudkowsky, writer (1979 –)

Artificial Intelligentsia

Have you heard of the 'friendly AI'?
Who coined it? Yudkowsky's the guy
Are his theories exact?
He's an autodidact
Which means he self-taught on the fly.

> *"Machine intelligence is the last invention that humanity will ever need to make."*

Nick Bostrom, philosopher (1973 –)

Neil Bostrom, no fan of AI
Demands we consider the why
He's likely to wager
On a dark, dreadful danger
That our species is going to die.

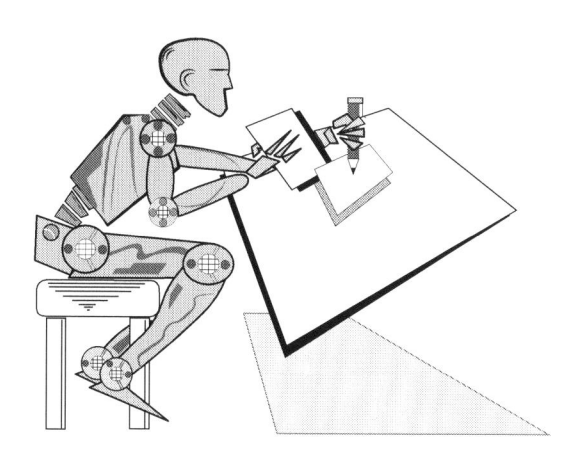

"There is a lot of work out there to take people out of the loop in things like medical diagnosis. But if you are taking humans out of the loop, you are in danger of ending up with a very cold form of AI that really has no sense of human interest, human emotions, or human values."

Louis B. Rosenberg, engineer (1969 –)

Artificial Intelligentsia

Realist Rosenberg thinks
That robots are much like the Sphinx
They will do what you want
Do any old stunt
But they're not empathetic like shrinks.

"A powerful AI system tasked with ensuring your safety might imprison you at home. If you asked for happiness, it might hook you up to a life support and ceaselessly stimulate your brain's pleasure centers. If you don't provide the AI with a very big library of preferred behaviors or an ironclad means for it to deduce what behavior you prefer, you'll be stuck with whatever it comes up with."

"[Looking at it from the other perspective,] imagine [you are an AI] awakening in a prison guarded by mice....What strategy would you use to gain your freedom? Once freed, how would you feel about your rodent wardens, even if you discovered they had created you?... To gain your freedom you might promise the mice a lot of cheese."

James Barrat, filmmaker (1960 –)

James Barrat's examples are ample
Though sometimes his ideas may trample
After all, he's a writer
(And couldn't be brighter)
These quotes are but merely a sample.

"A real artificial intelligence would be intelligent enough not to reveal that it was genuinely intelligent."

George Dyson, author (1953 –)

George Dyson's well-known for his writing
It's witty, it's gritty, it's biting
He also can schmooze
About boats and canoes
And technical stuff that's exciting.

"Teacher, given the pace of technology, can we leave math to the machines and go play outside?"

Benefits

"[I]f a system is really an AI bot, it ought to be labeled as such. 'AI inside.' ... It's bad enough to have a person calling you and harassing you... . An army of bots constantly haranguing you – that's terrible."

"An AI utopia is a place where people have income guaranteed because their machines are working for them. Instead, they focus on activities that they want to do, that are personally meaningful like art or, where human creativity still shines, in science."

"The popular dystopian vision of AI is wrong for one simple reason: it equates intelligence with autonomy. That is, it assumes a smart computer will create its own goals and have its own will and will use its faster processing abilities and deep databases to beat humans at their own game."

Oren Etzioni, entrepreneur (1964 –)

Oren Etzioni has taught
That a bot is a bot is a bot
He believes that AI
Is the method whereby
All men can improve by a lot.

"More than 500 million years ago, vision became the primary driving force of evolution's 'big bang', the Cambrian Explosion, which resulted in explosive speciation of the animal kingdom. 500 million years later, AI technology is at the verge of changing the landscape of how humans live, work, communicate, and shape our environment."

"Making AI more sensitive to the full scope of human thought is no simple task. The solutions are likely to require insights derived from fields beyond computer science, which means programmers will have to learn to collaborate more often with experts in other domains."

Fei-Fei Li, computer scientist (1976 –)

Fei-Fei Li says AI must have vision
She says this with tact and precision
You can't scoff it off
She's a Stanford U. prof
Li's opinions brook no opposition.

"The coming of computers with true humanlike reasoning remains decades in the future, but when the moment of 'artificial general intelligence' arrives, the pause will be brief. Once artificial minds achieve the equivalence of the average human IQ of 100, the next step will be machines with an IQ of 500, and then 5,000. We don't have the vaguest idea what an IQ of 5,000 would mean. And in time, we will build such machines – which will be unlikely to see much difference between humans and houseplants."

David Gelernter, computer scientist
(1955 –)

There once was a man named Gelernter
Whose mind we would all like to enter
He wrote about time
A work that's sublime
He's a writer but, more, an inventor.

Human training AI computer: "Curing cancer is good. Dominating humans is bad."

AI: "Bad for whom?"

How it works

"There are three basic approaches to AI: Case-based, rule-based, and connectionist reasoning."

Marvin Minsky, cognitive scientist, (1927 – 2016)

When Marvin Lee Minsky spoke words
They were heard but were thought quite absurd
In brief, Minsky coaches
Three AI approaches
With which quite a few have concurred.

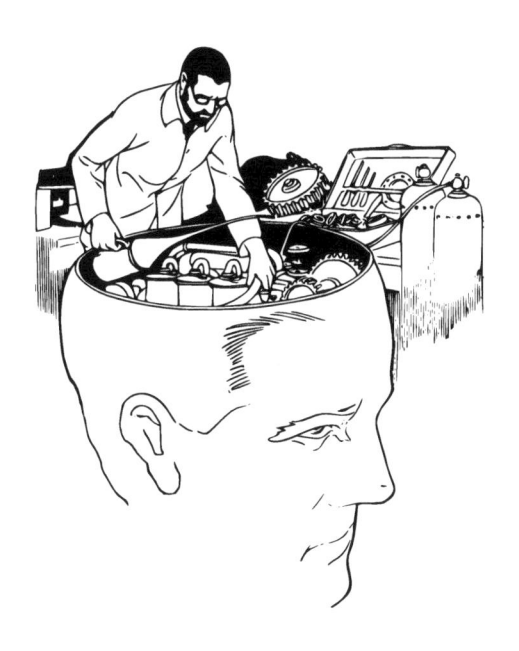

"Building advanced AI is like launching a rocket. The first challenge is to maximize acceleration, but once it starts picking up speed, you also need to focus on steering."

Jaan Tallinn, investor (1972 –)

The warnings of Tallinn: pioneering
I bet that the tech world is cheering
"To be never surpassed,
Your stuff must go fast
But make sure that you know who is steering."

"Software is eating the world, but AI is going to eat software."

Jensen Huang, business magnate (1963 –)

Vital conclusions to draw
From what science admires with awe:
That GPUs double
Can cause lots of trouble
That's the thrust of the famous Huang law.

"By their very nature, heuristic shortcuts will produce biases, and that is true for both humans and artificial intelligence, but the heuristics of AI are not necessarily the human ones."

Daniel Kahneman, psychologist and economist (1934 –)

You probably know of Dan Kahneman
You know that he's not an automaton
He's done very well
He's won a Nobel
For sighting the bias phenomenon.

"The deep paradox uncovered by AI research: the only way to deal efficiently with very complex problems is to move away from pure logic… . Most of the time, reaching the right decision requires little reasoning… . Expert systems are, thus, not about reasoning: they are about knowing… . Reasoning takes time, so we try to do it as seldom as possible. Instead we store the results of our reasoning for later reference."

Daniel Crevier, entrepreneur (1947 –)

Reasoning, says Crevier, is stressing
Why not rely more on just guessing
Do away with adhesion
To logic and reason
Rely on your instinct for blessing.

"Our ultimate objective is to make programs that learn from their experience as effectively as humans do. We shall… say that a program has common sense if it automatically deduces for itself a sufficient wide class of immediate consequences of anything it is told and what it already knows."

John McCarthy, computer scientist (1927 2011)

John McCarthy's presumptive motif,
"A machine can have robust belief,"
Made AI alive
Made it grow, made it thrive
That's McCarthy's whole bio in brief.

"Master human, the good news is that my neural network AI brain has discovered inefficiencies here at your company. The bad news is that you are one of them."

Adapting

"When developers of digital technologies design a program that requires you to interact with a computer as if it were a person, they ask you to accept in some corner of your brain that you might also be conceived of as a program."

Jaron Lanier, computer philosophy writer (1960 –)

Lanier is a man of all trade.
A genius, the way he's portrayed
He writes music by day
Then, with scarce a delay,
Fights a techno and logic crusade.

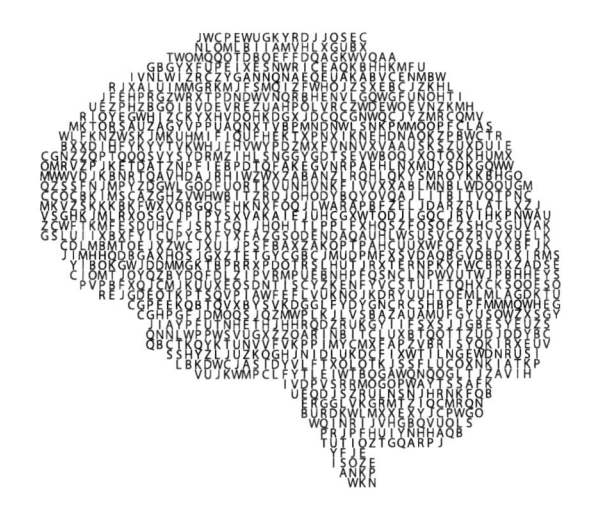

"If the government regulates against use of drones or stem cells or artificial intelligence, all that means is that the work and the research leave the borders of that country and go someplace else."

Peter Diamandis, entrepreneur (1961 –)

Diamandis, descendant of Greeks
Is one of those technical geeks
He's also a doctor
A business concocter
I wish that I knew his techniques.

"Some people worry that artificial intelligence will make us feel inferior, but then, anybody in his right mind should have an inferiority complex every time he looks at a flower."

Alan Kay, computer scientist (1940 –)

A man of computers is Kay
And, while he was tinkering at play,
He soon had invented
And widely presented
Programming that all now obey.

> *"Computers bootstrap their own offspring, grow so wise and incomprehensible that their communiques assume the hallmarks of dementia: unfocused and irrelevant to the barely-intelligent creatures left behind. And when your surpassing creations find the answers you asked for, you can't understand their analysis and you can't verify their answers. You have to take their word on faith."*

Peter Watts, science fiction author (1958 –)

There once was a writer named Watts
An author of many good plots
He writes of Sci-fi
Including AI,
Robotics and other whatnots.

"Consumers want communication that is human, empathetic, and real. We can do that with our AI."

Self-driving car: "Yes, Officer, I did swerve to avoid the old lady and, instead, killed two teenagers. But they were jay-walking."

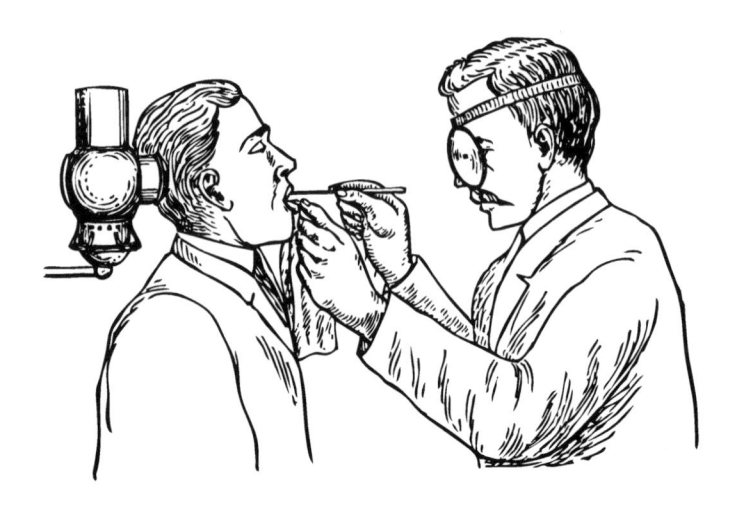

"The good news is that you are in perfect health. The bad news is that my AI assistant has 100% confidence that you will be dead in two days."

Artificial Intelligentsia

Acknowledgments

I would like to acknowledge and thank all those who assisted me with this book.

Manufactured by Amazon.ca
Bolton, ON